CENTRAL ASIA REGIONAL ECONOMIC COOPERATION PROGRAM

CLIMATE CHANGE ACTION PLAN

2025–2027

NOVEMBER 2024

Contents

Tables, Figure, and Boxes

Tables

Figure

Boxes

Acknowledgments

The Climate Change Action Plan was developed on behalf of the Central Asia Regional Economic Cooperation (CAREC) Secretariat and the Asian Development Bank (ADB) under the technical assistance Supporting Regional Actions to Address Climate Change under the CAREC Program. The team would like to thank Yevgeniy Zhukov, director general, Central and West Asia Department (CWRD); Xiaohong Yang, deputy director general, CWRD; Lyaziza Sabyrova, regional head, Regional Cooperation and Integration Unit, CWRD; Johannes Linn, nonresident senior fellow, The Brookings Institution, and ADB consultant; Irene de Roma, senior programs officer, CWRD; and Jennifer Lapis, project and events coordinator, for their invaluable support and guidance.

The development of the CAREC Climate Change Action Plan has been a collaborative effort involving numerous stakeholders dedicated to addressing the pressing challenges posed by climate change in the CAREC region. The team extends its heartfelt gratitude to the Working Group on Climate Change for their dedication and active participation in identifying key action plan areas and strategies that reflect the needs and priorities of member countries, while ensuring alignment with regional and global climate goals.

The team would like to thank CAREC sectors' and clusters' focal points for their engagement and valuable feedback. The team also appreciates the regional cooperation coordinators and national focal point advisors in all CAREC countries for their assistance in organizing and hosting in-person meetings and virtual consultations. The team extends its gratitude to senior officials in CAREC countries and development partners for their input, feedback, and contributions.

Together, these contributions have laid the groundwork for a comprehensive approach to climate adaptation and mitigation in the CAREC region, fostering a shared vision for a sustainable and resilient future.

The authors of the CAREC Climate Change Action Plan are Matteo Modè, senior climate change specialist, ADB consultant; Kristian Rosbach, senior regional cooperation specialist, CWRD; and Carmen Garcia Perez, regional cooperation specialist, CWRD.

Abbreviations

ADB	–	Asian Development Bank
AIIB	–	Asian Infrastructure Investment Bank
ASEAN	–	Association of Southeast Asian Nations
CAREC	–	Central Asia Regional Economic Cooperation
CCAP	–	CAREC Climate Change Action Plan
COP	–	Conference of the parties
CSPPF	–	Climate and Sustainability Project Preparatory Fund
G2F	–	glaciers-to-farms
GHG	–	greenhouse gas
GLOF	–	glacial lake outburst flood
MHEWS	–	multi-hazard early warning system
NBS	–	nature-based solution
NDC	–	nationally determined contribution
NFP	–	national focal point
SOM	–	Senior Officials' Meeting
WEF	–	water–energy–food
WGCC	–	Working Group on Climate Change

Glossary

Adaptation — In human systems, adaptation refers to the process of adjustment to actual or expected climate and its effects, to moderate harm or exploit beneficial opportunities. In natural systems, adaptation refers to the adjustment process to actual climate and its effects; human intervention may facilitate adjustment to expected climate and its effects.

Climate change — This refers to the long-term temperature shifts and weather patterns due to natural processes and human activities.

Disaster — Severe disruption of the functioning of a community or a society is triggered by geophysical or extreme weather hazard events leading to human, material, economic, or environmental losses and impacts. Disasters occur when hazard events and extreme weather hazard events interact with the exposure of vulnerable people and assets to those events.

National adaptation plans — The objectives of national adaptation plans are (i) to reduce vulnerability to the impacts of climate change by building adaptive capacity and resilience; and (ii) to integrate adaptation into new and existing national, sector, and subnational policies and programs, especially development strategies, plans, and budgets.

Nationally determined contributions — These embody the efforts that countries pledge to reduce their greenhouse gas emissions and adapt to the impacts of climate change. Nationally determined contributions (NDCs) are at the heart of the Paris Agreement and the achievement of its long-term goals. The Paris Agreement (Article 4, paragraph 2) requires each Party to prepare, communicate, and maintain successive NDCs that it intends to achieve. Parties shall pursue domestic mitigation measures to achieve the objectives of such contributions.

Natural climate solutions — These solutions apply specifically to climate mitigation and include energy efficiency, integration of renewable energy, sustainable transport and buildings, forest management, and sustainable agriculture. Nature-based solutions and natural climate solutions are sometimes used interchangeably depending on practitioners and areas of interventions.

Nature-based solutions — These are actions to protect, sustainably manage, or restore natural ecosystems that address societal challenges such as climate change, human health, food and water security, and disaster risk reduction effectively and adaptively, simultaneously providing human well-being and biodiversity benefits.

Paris Agreement — This agreement is a legally binding international treaty on climate change. It was adopted by 196 Parties at the 2015 United Nations Climate Change Conference (COP21). Its overarching goal is to hold "the increase in the global average temperature to well below 2°C above pre-industrial levels" and pursue efforts "to limit the temperature increase to 1.5°C above pre-industrial levels."

Resilience — This refers to the capacity of interconnected social, economic, and ecological systems to cope with a hazardous event, trend, or disturbance, responding or reorganizing in ways that maintain their essential function, identity, and structure. Resilience is a positive attribute when it maintains the capacity for adaptation, learning, and transformation.

Sources: Asian Development Bank. 2023. *Regional Action on Climate Change: A Vision for CAREC*; and United Nations Development Programme. 2023. *The Climate Dictionary*.

Executive Summary

The Central Asia Regional Economic Cooperation (CAREC) region is highly vulnerable to climate impacts, including rising temperatures, extreme weather events, and natural hazards, threatening energy and water availability, food security, and overall economic stability.

To combat these risks, significant regional investments are required to enhance resilience, develop climate-adaptive infrastructure, and support the transition toward low-carbon economic pathways, which not only align with global climate targets but also promote economic efficiency and job creation.

The CAREC Climate Change Action Plan

The CAREC Climate Change Action Plan (CCAP) builds upon the regional Climate Change Scoping Study and Regional Action on Climate Change: A Vision for CAREC (CAREC Climate Change Vision) endorsed at the 22nd CAREC Ministerial Conference in November 2023 in Tbilisi, Georgia. The CCAP provides a comprehensive framework to address climate change challenges and advance regional climate actions through enhanced coordination among the various CAREC sector committees and working groups.

A rolling 3-year plan spanning initially 2025–2027 and then replenished, the CCAP seeks to identify gaps in regional climate actions, prioritize and attract financing for regional climate adaptation and mitigation projects, and strengthen collaboration among development partners.

The CCAP was developed in close consultation with CAREC's Working Group on Climate Change, comprising representatives from member countries and development partners. It focuses on achieving a climate-resilient and low-carbon CAREC region through the following four thematic action areas:

(i) **Climate risk, preparedness, and health.** This area focuses on increasing the resilience of regional assets and enhancing the capacity of countries to prepare for and respond to climate and disaster risk. Initiatives include regional climate risk assessments and resilient infrastructure planning; a regional multi-hazard early warning system; health and climate change; and regional risk transfer solutions for the CAREC region.

(ii) **The water–energy–food security nexus.** This action area aims at increasing countries' resilience to climate shocks across the water, agriculture, and energy sectors. Initiatives include regional glacier risk assessment, water forecasting and climate adaptation in mountainous areas in collaboration with the Glaciers-to-Farms regional program; and irrigation, reservoir restoration, and river basin monitoring in Central Asia in coordination with the CAREC Water Pillar Working Group.

(iii) **Low-carbon growth.** This area focuses on mitigation actions and creating the enabling conditions to shift to a low-carbon economy. Areas of support will include decarbonization of trade and transport corridors, grid readiness and renewable energy integration, emissions reduction in urban areas, and carbon markets development.

(iv) **CAREC climate platform.** This area aims to establish a platform to conceptualize climate institutions for cooperation, facilitate capacity development, and propose financing sources for climate actions. Under this action area, support will be provided on capacity development, knowledge generation and dissemination, and climate finance and preparation of bankable regional climate projects.

The CCAP emphasizes gender equality and environmental sustainability, ensuring that climate actions consider the vulnerabilities faced by women and integrate safeguards to protect ecological resources. It also aligns with broader regional and national strategies and plans, enhancing cooperation among CAREC countries to effectively address climate impacts.

Implementation, Financing, and Monitoring

The implementation of the CCAP relies on an institutional framework involving multiple levels of governance. The CAREC Ministerial Conference will provide strategic guidance for the CCAP, meeting annually to discuss policies of regional relevance. At the same time, the CAREC Senior Officials' Meeting will monitor progress to ensure alignment with ministerial decisions and discuss regional climate adaptation and mitigation projects. The CAREC Working Group on Climate Change, with its country and development partner representatives, will facilitate the CCAP implementation, including integrating climate change across CAREC clusters in coordination with sector committees and working groups and promoting regional initiatives aligned with national climate strategies. The CAREC Secretariat will provide organizational support and coordinate efforts among member countries and development partners. The CAREC Climate Change Vision also anticipates the establishment of the CAREC Climate Change Steering Committee to oversee climate actions and align strategies across sectors, comprising CAREC national focal points and high-level officials from relevant ministries and agencies.

The practical implementation of the CCAP will require significant financial resources, as climate adaptation and mitigation have received limited investments in past CAREC projects. The CCAP will seek to mobilize resources through various approaches, including scaling up financial support from CAREC development partners and international climate funds, and developing innovative solutions to attract further investments from the private sector, including international reinsurance and capital markets for regional risk transfer solutions. Critical partners like the Asian Development Bank, the World Bank, and the Green Climate Fund are well positioned to lead financing efforts and support the implementation of the CCAP. Additionally, establishing the CAREC Climate and Sustainability Project Preparation Fund will assist CAREC countries in developing bankable regional climate projects aligned with their commitments under the Paris Agreement.

Mountainous area near Karakol, Kyrgyz Republic. Mountainous areas in the CAREC region are among the most vulnerable to climate change, experiencing increased occurrence and severity of extreme weather events and natural hazards, such as glacial lake outburst floods and landslides (photo by Matteo Modè).

1 Background

The Central Asia Regional Economic Cooperation (CAREC) Program is a partnership of 11 countries and development partners that cooperate for mutual benefit and greater regional prosperity.[1] CAREC countries are highly vulnerable to climate change impacts and natural hazards,[2] which are set to increase in the future.[3] Climate change projections show that, on average, the CAREC region is set to become considerably warmer, leading to a shift in climatic patterns, extreme rainfall events, heat waves, droughts, and glacier retreats.[4] These climate impacts will have compounding effects on energy and water availability, food security, and the people and economies of the region.

Climate impacts cause fatalities and displacement, disrupt essential services, and damage infrastructure and the built environment, with repercussions on critical sectors such as energy, water, transport, and health. From 2000 to 2024, floods, the most prominent natural hazard in the region, caused more than $1 trillion in losses.[5] Although statistically less significant, droughts are proportionally more damaging and have caused $370 billion in losses.[6]

During March and April 2024, northern Kazakhstan experienced its worst floods in 80 years. The floods affected 18,000 households, caused the displacement of 118,200 people, and led the Government of Kazakhstan to declare a state of emergency in 10 of the country's 17 regions.[7]

---

[1] CAREC countries include Afghanistan, Azerbaijan, the People's Republic of China, Georgia, Kazakhstan, the Kyrgyz Republic, Mongolia, Pakistan, Tajikistan, Turkmenistan, and Uzbekistan. ADB placed its regular assistance to Afghanistan on hold effective 15 August 2021..

[2] ADB. 2023. *CAREC 2030: Supporting Regional Actions to Address Climate Change—A Scoping Study.*

[3] Most CAREC countries rank with high or very high vulnerability to climate change and natural hazards, according to the Notre Dame Global Adaptation Initiative (ND-GAIN) Index, which assesses countries' vulnerability to climate change and other global challenges, in combination with their readiness to improve resilience to climate change.

[4] Projections according to the World Bank's Climate Change Knowledge Portal (accessed August 2024) show that the region will be 2°C warmer by the 2050s and more than 3°C warmer by the 2090s under conservative emission scenarios, while high emissions scenarios project an average increase in excess of 5°C by the 2090s.

[5] The Centre for Research on the Epidemiology of Disasters' EM-DAT International Disaster Database shows that an estimated 66% of natural hazards in the region are caused by floods (accessed July 2024).

[6] Centre for Research on the Epidemiology of Disasters. EM-DAT International Disaster Database (accessed July 2024).

[7] ReliefWeb. Kazakhstan: Floods - Mar 2024.

In August 2022, Pakistan experienced catastrophic floods triggered by unprecedented rainfall following a prolonged drought. Due to the compounding climate shocks, the floods killed 1,739 people and affected 2.3 million people, with an estimated $14.9 billion of economic losses.[8] During the winter of 2023–2024, Mongolia faced its second most severe *dzud* since 1945, and the fourth one in a decade.[9] The associated impacts resulted in the loss of 7.8 million livestock (13% of the country's total), with an estimated economic value of MNT2.5 trillion (approximately $733.5 million). The 1998 floods in the People's Republic of China affected 223 million people, claimed the lives of 4,150, and caused direct economic losses of $35 billion (in 2018 values).[10]

Mountainous areas are among the most vulnerable in the CAREC region because climate impacts such as temperature increases and extreme and unseasonal weather events are felt here most strongly.[11] Mountainous areas are increasingly experiencing glacier retreats, flash floods, glacial lake outburst floods (GLOFs), landslides, and other associated natural hazards, with wider repercussions on downstream water availability, energy production, food security and livelihoods of local communities, increased disaster risk, and human mobility.[12]

Heat stress and extremely high temperatures are also increasing throughout the region, particularly in lowlands, posing energy, water, and transport infrastructure challenges.[13] Exposure to extreme heat increases the risk of morbidity and mortality from several cardiovascular and respiratory diseases, especially among the most vulnerable, such as children, women, and older people. Between 2000 and 2019, approximately 489,000 heat-related deaths occurred each year globally, with 45% of these in Asia and 36% in Europe.[14]

To overcome these challenges, significant regional investments are required to increase the CAREC region's resilience to climate shocks and ensure that energy, water, and transport infrastructure is climate-resilient. Investments in climate adaptation measures will be needed to increase the preparedness and readiness of countries and to reduce future damage and losses, shifting to early warning and preemptive disaster risk management, while also leveraging resources to finance adaptation projects, including risk insurance instruments, as part of a comprehensive disaster risk financing strategy.

CAREC countries have opportunities to continue shifting to low-carbon economic pathways. This would contribute to national and global climate targets to hold warming to well below 2°C but also improve economic efficiency of energy use and reduce environmental damage, especially

---

[8]   Government of Pakistan, Ministry of Planning Development and Special Initiatives. 2022. *Pakistan Floods 2022: Post-Disaster Needs Assessment—Main Report.*

[9]   A *dzud* is a multifaceted natural hazard event occurring in Mongolia due to a cold dry climate, and encompassing drought, heavy snowfall, extreme cold, and windstorms. It impacts Mongolia's social and economic sectors as well as many other key development issues such as public health, migration, urbanization, unemployment, and livelihoods.

[10]   *Reuters.* 2024. China's Rains and Floods Led to Near Doubling of Natural Disaster Losses in July. 9 August.

[11]   S. Manandhar et al. 2018. Climate Vulnerability and Adaptive Capacity of Mountain Societies in Central Asia. *Mountain Societies Research Institute Research Paper.* No. 3. University of Central Asia.

[12]   Z. Saidaliyeva et al. 2024. Adaptation to Climate Change in the Mountain Regions of Central Asia: A Systematic Literature Review. *Wiley Interdisciplinary Reviews Climate Change.* 15 (5).

[13]   Future climate projections are based on the Coupled Model Intercomparison Project Phase 6 models, which are utilized within the *Sixth Assessment Report of the Intergovernmental Panel on Climate Change,* providing estimates of future temperatures and precipitation. Future scenarios are based on different assumed levels of greenhouse gas (GHG) emissions.

[14]   Q. Zhao et al. 2021. Global, Regional, and National Burden of Mortality Associated with Non-Optimal Ambient Temperatures from 2000 to 2019: A Three-Stage Modelling Study. *Lancet Planet Health.* 5 (7).

pollution. Moreover, it would generate new jobs and economic opportunities via decarbonization, integration of renewable energy into the regional grid, and integration of innovative climate-smart solutions in trade and agriculture, among other areas.

Climate mitigation and adaptation actions should be pursued at regional scale since climate change impacts are, by their own nature, regional and transboundary. For example, glacier melt affects the availability of water and energy resources in the region, expanded reliance on renewable energy requires strengthening the regional energy grid (including energy storage and hydropower integration), and natural hazards such as droughts and floods may have regionwide impacts and need regionally coordinated weather and climate prediction and early warning actions. Regional transport networks are best decarbonized in a coordinated manner (electrification, public transport integration), and technology, knowledge, and best practices can be effectively shared on a regional basis (footnote 2).

The CAREC 2030 Strategy, approved in 2017, provides the long-term strategic framework for the CAREC Program leading to 2030. The strategy refers to climate change as a regional challenge but does not explicitly include climate action as a crosscutting focal area. Despite this, CAREC has rapidly adapted and taken several steps to respond to member countries' changing needs and priorities. CAREC is committed to addressing the pressing challenges of climate change and has been focusing on strengthening the resilience of its operations and activities to climate impacts, while also contributing to global efforts to mitigate greenhouse gas (GHG) emissions and achieve the goals of the Paris Agreement.[15]

In 2023, the CAREC Secretariat conducted a detailed regional climate change scoping study, which highlighted the urgent need to support its member countries in reinforcing, modifying, and implementing existing national strategies on climate change mitigation and adaptation and in developing a range of regional actions in response to climate change impacts and required solutions (footnote 2). Among identified actions, the scoping study recommended preparing a CAREC climate change strategy for adoption by CAREC ministers, and a road map for climate mitigation and adaptation projects to be designed, implemented, and financed under the CAREC Program.

Building on the CAREC climate change scoping study, a regional action on climate change vision was endorsed at the 22nd CAREC Ministerial Conference in November 2023 in Tbilisi, Georgia.[16] This regional vision has three broad goals: climate adaptation, mitigation, and regional transboundary cooperation on climate change. It is based on five principles: (i) aligning with national adaptation strategies and with the Paris Agreement, (ii) deepening regional cooperation on climate change, (iii) expanding coordination with all development partners on regional climate action, (iv) integrating the role of the private sector and civil society into the regional climate dialogue, and (v) building an open and inclusive regional institutional platform on climate change.

---

[15] The Paris Agreement is a legally binding international treaty on climate change adopted by 196 parties at the UN Climate Change Conference (COP21) in Paris, France, on 12 December 2015. It entered into force on 4 November 2016. For details on the agreement's goals, see United Nations Framework Convention on Climate Change (UNFCCC). Key Aspects of the Paris Agreement.

[16] ADB. 2024. Regional Action on Climate Change: A Vision for the Central Asia Regional Economic Cooperation Program.

Kyrgyz pastoralists along the Bishkek–Osh road corridor. As mountainous areas are most vulnerable to climate change, the Climate Change Action Plan identifies required adaptation measures for key sectors and communities (photo by Matteo Modè).

2 The CAREC Climate Change Action Plan

The CAREC Climate Change Action Plan (CCAP) implements key steps and recommendations identified by the CAREC regional action on climate change vision.[17] The action plan is a rolling plan spanning 3 years, starting 2025–2027 and then replenished, seeking to (i) coordinate regional climate actions in the CAREC clusters and working groups, (ii) help prioritize regional climate adaptation and mitigation projects and initiatives across CAREC sectors, and (iii) strengthen coordination among development partners to increase and optimize resources to support regional climate actions. The CCAP aims to identify gaps in regional climate actions and attract financing to develop and implement regional investment projects.

The CCAP was developed in close consultation and coordination with members of the CAREC Working Group on Climate Change (WGCC). This involved a participatory planning process where CAREC member countries and development partners identified action plan areas and potential activities.[18] The formulation of the action plan was also coordinated with various sector committees, working groups, and focal points across CAREC clusters. It is informed by CAREC's project portfolio and countries' national adaptation priorities[19] and nationally determined contributions (NDCs).[20]

The outcome of the CCAP is a climate-resilient and low-carbon CAREC region. The CCAP focuses on four thematic action plan areas (as illustrated in the succeeding figure).

---

[17] The CCAP development is supported by the ADB regional technical assistance, Supporting Regional Actions to Address Climate Change under the CAREC Program.

[18] CAREC WGCC. 2024. First Meeting of the Central Asia Regional Economic Cooperation Working Group on Climate Change. Summary of Discussions.

[19] Concrete proposals for regional adaptation projects were raised by member countries during the CCAP consultative process according to their respective resilience and adaptation needs, taking into consideration that most CAREC countries are still in the process of developing their national adaptation plans. The status of plan submissions can be found at the UNFCCC's National Action Plan Central.

[20] An analysis of NDCs' areas of intervention, targets, key priority actions, and means of implementation was conducted in relation to the CCAP outcomes, outputs, and identified regional initiatives across the four action plan areas.

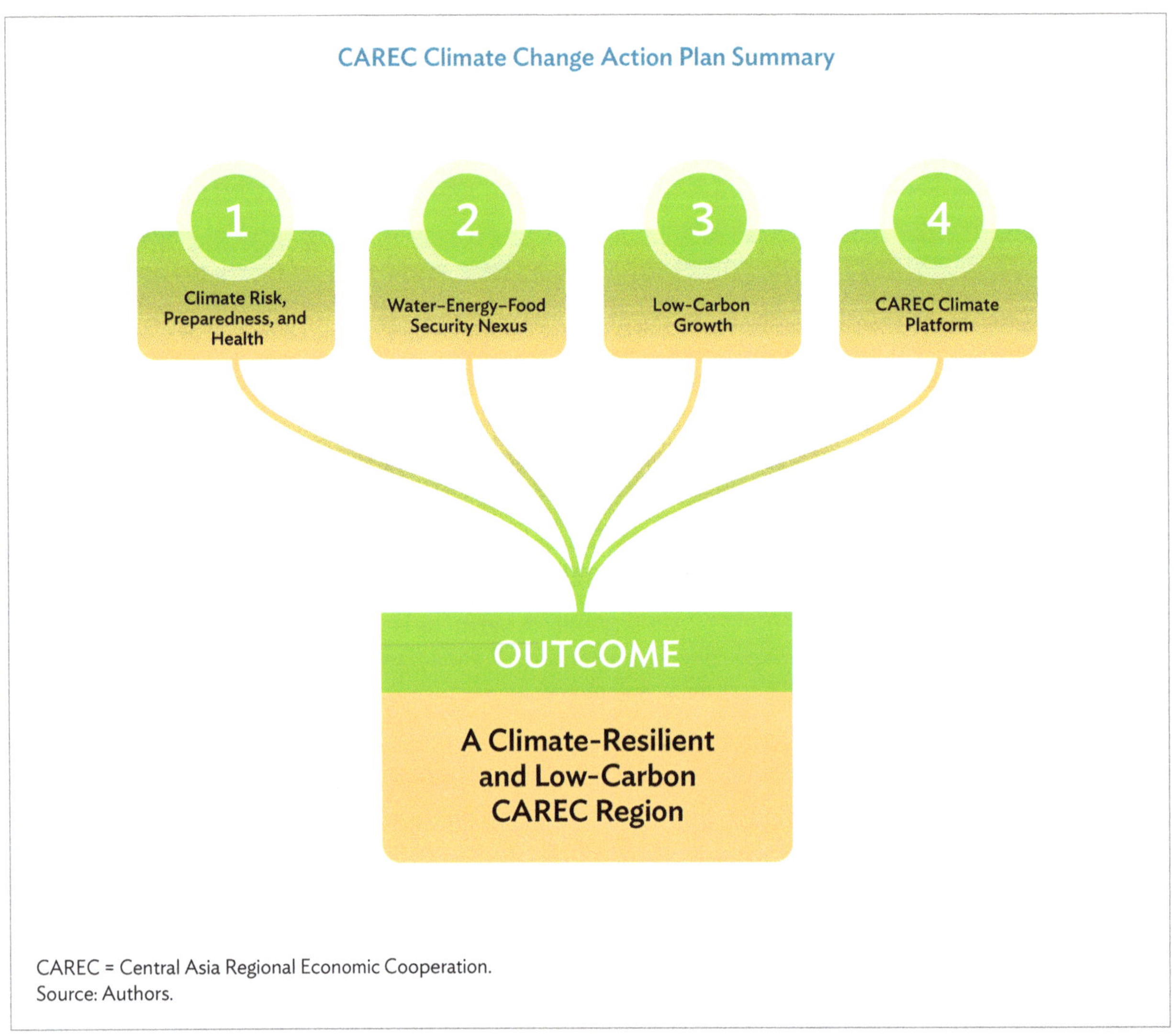

CAREC = Central Asia Regional Economic Cooperation.
Source: Authors.

The first action plan area, **Climate Risk, Preparedness, and Health**, focuses on safeguarding communities, economies, and infrastructure by increasing the resilience of regional assets and enhancing countries' capacity to prepare for and respond to climate and disaster risk. The action plan area includes regional climate risk assessments and resilient infrastructure planning, a regional multi-hazard early warning system (MHEWS), health and climate change, and risk transfer solutions for the CAREC region.

The second action plan area, **Water–Energy–Food Security Nexus**, is on regional water forecasting and water projects, the resilience of mountainous regions, and climate-smart agriculture. Initiatives in this action area are (i) regional glacier risk assessment and water forecasting—preparatory assessments and supporting actions in collaboration with the Glaciers-to-Farms (G2F) program of the Asian Development Bank (ADB); (ii) irrigation, reservoir restoration, and river basin monitoring in Central Asia (in coordination with the CAREC Water Pillar); and (iii) climate adaptation in mountainous areas (in collaboration with the G2F program).

The third action plan area, **Low-Carbon Growth**, is on mitigation actions focusing on decarbonizing trade and transport corridors, increasing grid readiness, and reducing emissions in urban areas. Low-carbon growth will be achieved through low-emission zones, carbon market development, decarbonization of CAREC transport and trade corridors, and grid readiness and renewable energy integration.

The fourth action plan area, **CAREC Climate Platform**, conceptualizes climate institutions for cooperation, facilitates capacity development, and proposes financing sources for climate actions. This action plan area has the following subcategories: (i) capacity development on regional climate action, (ii) regional platform for knowledge generation and dissemination, and (iii) climate finance and development of bankable regional climate projects.

The CCAP reflects gender aspects in the formulation and implementation of climate activities, taking into consideration the CAREC Gender Strategy 2030.[21] Climate change affects women and girls disproportionately and further increases gender inequality. Research, for instance, indicates that women, especially in developing countries, are disproportionately affected by heat waves.[22] When formulating and implementing climate change policies, strategies, and actions, it is essential to fully consider the specific needs of women and girls and ensure women's full and equal participation. Gender equality concerns will be integrated across action plan areas, regional initiatives, and projects. This will ensure that gender disparities are taken into consideration in the development and implementation of resilience measures as well as in low-carbon just transition activities to help reduce the vulnerability of women and girls to climate shocks and ensure equitable benefits.

Rice farming in Ferghana Valley, near Andijan, Uzbekistan. Agriculture is among the sectors that will be affected by increasing water scarcity in the future (photo by Matteo Modè).

21   ADB. 2021. *CAREC Gender Strategy 2030: Inclusion, Empowerment, and Resilience for All*.
22   ADB. Triple Threat: The Health, Social, and Economic Impacts of Heat Waves on Women. *Rising Above the Heat: Strengthening Women's Resilience to Heat Stress* (multimedia).

The CCAP also integrates environmental sustainability across action plan areas and regional initiatives to ensure that adaptation and resilience measures include the required environmental safeguards to limit harmful and unintended impacts on environmental receptors during implementation of activities and projects.

The CCAP has solid linkages and synergies with other regional climate strategies. This includes ADB's forthcoming Climate Change Road Map in Central and West Asia (2024–2030),[23] which identifies climate priorities for the bank's operations in alignment with its Climate Change Action Plan.[24] The road map puts forward recommendations and strategic directions aimed at raising climate actions, investments, and finance in the region, noting the role of the CAREC platform to help advance potential regional climate projects and cooperation initiatives identified in the road map.

In addition, there are linkages with the Regional Climate Change Adaptation Strategy in Central Asia (2023–2030), which was developed and approved by Kazakhstan, the Kyrgyz Republic, Tajikistan, Turkmenistan, and Uzbekistan.[25] The overall goal of this regional strategy is to develop a mechanism for cooperation between Central Asian countries to overcome climate impacts and implement adaptation measures, therefore synergizing with the CCAP.

---

[23]   ADB's Climate Change Road Map in Central and West Asia covers 10 countries: Afghanistan, Armenia, Azerbaijan, Georgia, Kazakhstan, the Kyrgyz Republic, Pakistan, Tajikistan, Turkmenistan, and Uzbekistan.

[24]   ADB. 2023. *Climate Change Action Plan, 2023–2030*.

[25]   Supported by Germany's development agency, Deutsche Gesellschaft für Internationale Zusammenarbeit (GIZ), with the Regional Environmental Centre for Central Asia (CAREC Eco). The strategy was presented at the 28th Conference of the Parties to the UNFCCC in Dubai, 2023 (COP28). See CAREC Eco. 2023. *Regional Climate Change Adaptation Strategy for Central Asia (2023–2030)*. GIZ.

3 Action Plan Areas

The action plan areas of intervention build upon the strategic recommendations of the CAREC Climate Change Vision, with a primary focus on adaptation as proposed by the WGCC during the formulation process. In addition, implementing the CCAP adaptation and mitigation activities under the first three action plan areas is supported by the establishment of a CAREC climate platform (action plan area four) to allow for the required transboundary cooperation, coordination, and knowledge sharing on climate actions.

Integrating climate change into CAREC sectors, clusters, and investments. To reduce overlap and maximize available climate finance resources, the CCAP will not seek to implement all the activities identified by the WGCC directly and included in the action plan areas. Instead, the CCAP builds upon existing and planned activities, some of which are being and/or will be implemented by CAREC sector working groups and development partners. The value added by the CCAP, in line with CAREC's overall mandate, is to provide the required regional coordination and knowledge exchange via the WGCC and the CAREC climate platform on several climate actions. The proposed regional activities will be implemented in close coordination with CAREC sector committees and working groups and integrated into their respective strategies and work plans as applicable.

The CCAP endeavors to prioritize and sequence potential regional climate actions that could be implemented and/or initiated during 2025 to 2027. However, not all proposed interventions and actions will be completed during the 3-year period. The implementation progress of the CCAP will be reviewed annually by the WGCC, and follow-up action plans will be developed (over 3-year cycles), considering the main results achieved and lessons learned from the CCAP implementation, as well as the changing needs and priorities from member countries.

Climate Risk, Preparedness, and Health

Overview

This action plan area focuses on safeguarding communities, economies, and infrastructure by increasing the resilience of regional assets and enhancing countries' capacity to prepare for and respond to climate and disaster risks. This involves comprehensive regional climate risk assessments, which are vital for understanding the potential impacts of climate change; identifying the most vulnerable sectors, communities, and assets; and allowing climate finance to be targeted effectively toward the most urgent and high-impact regional interventions (Table 1).

Table 1: Action Plan Area 1—Climate Risk, Preparedness, and Health

OBJECTIVE		
To increase resilience of regional infrastructure and communities, and countries' capacities to prepare for and respond to climate and disaster risk		
Regional Initiative	**Coordinating Country**[a]	**Indicative Start Date**[b]
Regional climate risk assessments and resilient infrastructure planning	Kazakhstan, Pakistan	2024
Health and climate change	People's Republic of China, Pakistan	2025
Risk transfer solutions for the CAREC region	Kyrgyz Republic, Tajikistan	2025
Regional multi-hazard early warning system	Azerbaijan, Mongolia, Uzbekistan	2026

CAREC = Central Asia Regional Economic Cooperation.

[a] Responsible for coordinating the implementation of the regional initiatives and projects, and reporting on progress during working group meetings to the CAREC Climate Change Steering Committee, once established.
[b] The year 2024 is the preparatory phase of the Climate Change Action Plan, including the initiation of some pilot actions in various action plan areas.

Source: Asian Development Bank.

Regional Climate Risk Assessments and Resilient Infrastructure Planning

This regional initiative involves conducting comprehensive assessments of present and projected climate risks facing the CAREC region, focusing on critical regional assets to identify hot spot areas and sectors most vulnerable to floods, landslides, mudflows, and extreme heat. Historical climate trends and future climate projections for the region will be analyzed based on the latest climate science and models (e.g., CMIP6).[26] A climate vulnerability map of regional CAREC

---

[26] CMIP6 is the Coupled Model Intercomparison Project Phase 6, 2016–2021 of the Intergovernmental Panel for Climate Change (IPCC) Sixth Assessment Report, which uses climate scenarios such as the Shared Socioeconomic Pathways to allow climate change practitioners to develop projected socioeconomic global changes according to projected conservative, moderate, and high emissions levels.

Pakistan Disaster Needs Assessment Project, 2010. This ADB project constructed a third emergency bridge across the Kacchi Canal in Muzaffargarh District, Punjab Province (photo by ADB).

assets could be developed. Potentially vulnerable CAREC assets include transport corridors (road, railways, border crossing points); and water, energy, and irrigation infrastructure such as dams and hydropower plants, water storage basins and reservoirs, and irrigation canals.[27]

These assessments will allow the identification of opportunities and priority areas for adaptation actions and investment projects to increase the resilience of regional assets. This may include, for example, updating road construction standards to increase resilience and reduce vulnerability of critical assets based on a regional review of national standards and construction rules and regulations.

Under this initiative, CAREC will support technical and institutional capacity building to conduct regional climate risk assessments and exchange knowledge, best practices, and innovative climate solutions among countries. Collaboration with CAREC sector committees and working groups will be strengthened to conduct climate risk assessments and to integrate the findings into regional strategies and sector adaptation plans.

Regional Multi-Hazard Early Warning System

This regional initiative aims to establish a regional multi-hazard early warning system (MHEWS) to strengthen disaster resilience in the CAREC region and provide a coordinated and harmonized approach to climate change risk monitoring, data sharing, and early action. The main objective of the MHEWS is to strengthen regional cooperation and information sharing on climate risks and hazards by developing a shared transboundary data platform and knowledge management system.

---

[27] A checklist and terms of reference for the climate risk assessment have already been developed based on standard climate risk and vulnerability assessment practice, including a desk review of available data, field assessment, and interviews and discussions with relevant stakeholders.

Managing risk at Sarez Lake in Rushon District, Gorno-Badakhshan Autonomous Region (GBAO), Tajikistan. A potential outburst of the lake could affect 5 million people living in four different countries (Afghanistan, Tajikistan, Turkmenistan, and Uzbekistan). Managing its risk via an early warning system is crucial to protect communities downstream of the dam (photo by Matteo Modè).

This initiative will support the development of common guidelines and early warning protocols for coordinated action and post-disaster recovery efforts, considering existing national early warning systems and resources. Support will include capacity building for hydrometeorological services; and the modernization, integration, and harmonization of climate and disaster risk monitoring infrastructure, including early warning equipment, communication networks, and forecasting, taking into consideration health risks.

The natural hazards selected for monitoring will depend on the scope of the MHEWS, which will be jointly defined by CAREC members during the formulation of this initiative. In addition to potentially covering floods and hydrogeological hazards, the MHEWS will seek to integrate other hazards, such as heat waves and droughts. Further cooperation among CAREC member countries will be considered, taking into account existing initiatives by CAREC development partners, such as the United Nations Development Programme and the World Bank.[28]

The MHEWS will enhance CAREC countries' capacities for risk assessment, early warning, and emergency response planning, promoting the engagement of local communities and including gender aspects in the design of relevant components to ensure its relevance and accessibility within national and local contexts. Establishment of the system will also include the promotion

---

[28] Hosting of the MHEWS is subject to discussion with member countries, with Azerbaijan and Uzbekistan initially expressing interest in being host. The United Nations Development Programme is implementing a project, Enhancing Multi-Hazard Early Warning System to Increase Resilience of Uzbekistan Communities to Climate Change Induced Hazards, upon which CAREC can build. The World Bank meanwhile has its Central Asia Hydrometeorology Modernization initiative, which aims to strengthen hydrological and meteorological services and foster regional coordination and knowledge exchange. Both provide opportunities for coordination and collaboration.

of public awareness and education on disaster risk reduction and appropriate response actions.[29] Capacity building and technical assistance on early warning systems management and integration to disaster management and hydrometeorological agencies will also be provided.

Health and Climate Change

This initiative integrates heat risks in the regional MHEWS to respond to increasing heat risk and its impact on health, particularly on the most vulnerable groups. Under conservative and high emissions scenarios, the annual probability of heat waves could increase significantly by the 2050s and even more markedly by the 2090s, particularly in the region's lowlands.[30] Exposure to extremely high temperatures and heat stress are among the leading causes of death globally, particularly among the most vulnerable, and can exacerbate underlying conditions such as cardiovascular and respiratory diseases, as well as causing stillbirths and miscarriages.[31]

Preparedness for extreme temperatures, including heat waves and cold waves, is an emerging area, both in the region and globally, with considerable gaps in awareness and response. Technical support will be provided to facilitate the integration of heat risk in the MHEWS to strengthen the climate capacities of health authorities for heat-wave forecasting, alerts, preparedness, and appropriate responses, with a focus on critical geographic areas and vulnerable groups. This could include capacity-building activities to raise climate risk awareness and heat-wave preparedness of medical staff and health facility managers, including addressing the reproductive health needs of women (e.g., antenatal and postnatal care), as part of an integrated disaster preparedness approach.

Other potential activities could include building climate-resilient health systems and upgrading health facilities in border areas via climate-resilient and energy-efficient measures, including knowledge sharing on green building standards, leading to greenhouse gas (GHG) savings and mitigation actions aligned with the Paris Agreement goals and countries' NDCs.[32] The CAREC climate platform will provide the required venue for regional training, exchange of experiences, and sharing of best practices on heat risk and health, including cross-sector activities in collaboration with the CAREC health working group.[33]

Risk Transfer Solutions for the CAREC Region

This initiative aims to improve countries' capacity to acquire disaster risk financing and enhance their public sector budget resilience against disaster events through regional collaboration, leveraging the international insurance and capital markets. The proposed regional project will build on an ADB-funded technical assistance, Developing a Disaster Risk Transfer Facility in the CAREC Region, as well as ongoing regional disaster risk management and financing initiatives.[34]

---

[29]   Integration and data exchange with other sectors such as health could be considered to increase monitoring of climate-sensitive diseases and prepare timely responses for the health sector to mitigate adverse health outcomes.

[30]   World Bank. Climate Change Knowledge Portal (accessed August 2024).

[31]   World Health Organization. Heatwaves.

[32]   ADB's experience in the region includes providing technical assistance in Tajikistan for the Maternal and Child Health Integrated Care Project. This project incorporates a climate risk assessment of maternal and child health facilities aimed at identifying climate-resilient and energy-efficient measures to be integrated into their design.

[33]   The CAREC Regional Investment Framework for Health 2022–2027 has been updated to include actions on climate and health.

[34]   ADB. Regional: Developing a Disaster Risk Transfer Facility in the Central Asia Regional Economic Cooperation Region.

Impact of mudslides on transport infrastructure. A mudslide badly damaged a road in the town of Vakhdat, Tajikistan (photo by Tajikistan Committee for Emergency Situations and Civil Defence).

It also responds to the joint statement endorsed at the 22nd CAREC Ministerial Conference in 2023, in which CAREC countries requested support to further advance the implementation of regional risk transfer solutions.[35]

This regional initiative will support further development of the long-term ambition to establish a CAREC risk transfer facility, based on experiences from existing regional risk pools in Africa, the Caribbean and Central America, Pacific island countries, and Southeast Asia.[36] A CAREC risk transfer facility will complement national disaster risk management initiatives while providing additional benefits such as broader risk diversification, lower operational costs, greater access to international insurance and capital markets, and a coordinated risk analysis and modeling approach. The proposed initiative will also support the implementation of pilot risk financing instruments (e.g., disaster relief bonds) for selected CAREC countries.[37]

Capacity building and knowledge support on climate and disaster risk modelling and financing will be provided, taking into consideration countries' varying levels of capacity. Regional workshops with public and private stakeholders will be conducted to promote the exchange of knowledge, best practices, and lessons learned from other regions.[38]

Linkages and Synergies with Existing Initiatives

The proposed interventions under action plan area 1 can build upon and establish linkages with past and ongoing global and regional initiatives from development partners, as well as existing CAREC institutional structures (Table 2).

---

[35]   CAREC. 2023. *22nd CAREC Ministerial Conference Joint Ministerial Statement*. Tbilisi, Georgia. 30 November.
[36]   ADB. 2023. *Road Map to Developing a Regional Risk Transfer Facility for CAREC*.
[37]   The initial focus of the pilot disaster relief bond will be selected countries (most likely countries eligible to receive Asian Development Fund support), although all CAREC countries will be observers to the pilot issuance process.
[38]   This includes local insurance industry stakeholders and insurance supervisors, as well as disaster risk management agencies, the Ministry of Finance, and other relevant government authorities.

Table 2: Linkages and Synergies for Climate Risk, Preparedness, and Health

Regional Initiative	Project Particulars
Regional Climate Risk Assessments and Resilient Infrastructure Planning	▪ Coordination with the CAREC Transport Sector Committee for climate risk assessment of regional transport and cross-border infrastructure to integrate climate resilience measures ▪ Country and project-based climate risk assessments conducted by CAREC development partners ▪ Pilot project, Kashkadarya Masterplan Risk Screening, whose approach and methodology could be replicated at the regional level[a]
Regional Multi-Hazard Early Warning System	▪ The United Nations Early Warnings for All Initiative, formally launched in 2022 at the 27th Conference of the Parties to the United Nations Framework Convention on Climate Change in Sharm El-Sheikh (COP27 meeting) and jointly endorsed by multilateral development banks and development partners, calls for everyone to be protected by early warning systems by 2027.[b] ▪ Opportunities for cooperation with the Systematic Observations Financing Facility could be a source of climate finance for the CAREC MHEWS. The facility is one of the main pillars of the Early Warnings for All Initiative and funds improvements in weather and climate observation infrastructure in least-developed countries and small independent states, and analytical work in low- and middle-income countries.[c] ▪ The World Bank, Global Facility for Disaster Reduction and Recovery, and World Meteorological Organization joint initiative, the Central Asia Flood Early Warning System, is part of the ongoing work on improving hydrometeorological service delivery in the region.[d] ▪ The Global Telecommunication System, a key component within the World Meteorological Organization Information System, aims to facilitate the flow of data and processed products to meet World Weather Watch requirements.[e] ▪ ADB's regional project, Increasing Investments in Early Warning Systems to Strengthen Climate and Disaster Resilience, includes CAREC countries Kazakhstan, the Kyrgyz Republic, Mongolia, Tajikistan, and Uzbekistan.[f] ▪ The United Nations Development Programme's project for Enhancing Multi-Hazard Early Warning System to Increase Resilience of Uzbekistan Communities to Climate Change Induced Hazards (financed by the Green Climate Fund)[g] ▪ ADB and World Bank collaboration and data sharing on early warning for projects in Central Asia
Health and Climate Change	▪ ADB's project for Building a Climate Change Early Warning System for the Aged in the People's Republic of China aims to develop a climate change early warning system to warn the vulnerable older population and government authorities about heat risk. ▪ Collaboration with the CAREC Health Working Group is needed to strengthen health authorities' climate risk awareness and heat-wave preparedness in CAREC countries and support the integration of heat risk in MHEWS.
Regional Risk Transfer Solutions for the CAREC Region	▪ ADB-funded regional technical assistance, Developing a Disaster Risk Transfer Facility in the CAREC Region ▪ Strengthening Financial Resilience and Accelerating Risk Reduction in Central Asia Program is supported by the European Union, Global Facility for Disaster Reduction and Recovery, and the World Bank.

ADB = Asian Development Bank, CAREC = Central Asia Regional Economic Cooperation, MHEWS = multi-hazard early warning system.

[a] ADB Regional: Supporting Adaptation Decision Making for Climate Resilient Investments. The technical assistance aims to improve the understanding of developing member countries about how to make effective use of climate information and services to facilitate planning and decision-making under climate uncertainty and to generate, interpret, and apply climate information in decision-making in sectors including agriculture, water, and energy.

[b] United Nations. Early Warnings for All.

[c] The Systematic Observations Financing Facility provides grant funding for the collection of weather and climate data.

[d] World Bank. 2021. Central Asian Flood Early Warning System. Infographics. 10 December.

[e] World Meteorological Organization. Global Telecommunication System (GTS).

[f] ADB. Regional: Increasing Investments in Early Warning Systems to Strengthen Climate and Disaster Resilience.

[g] United Nations Development Programme, Climate Change Adaptation. Enhancing Multi-Hazard Early Warning System to Increase Resilience of Uzbekistan Communities to Climate Change Induced Hazards.

Source: ADB.

Water–Energy–Food Security Nexus

Overview

The CAREC region is heavily dependent on its glacier-fed water resources. As such, it faces significant challenges at the intersection of water, energy, and food security. Table 3 summarizes the regional initiatives and projects under the water–energy–food (WEF) security nexus (Action Plan Area 2).

Table 3: Action Plan Area 2—Water–Energy–Food Security Nexus

OBJECTIVE		
To increase countries' financial resilience to climate shocks across the water, energy, and agriculture sectors		
Regional Initiative	**Coordinating Country[a]**	**Indicative Start Date**
Regional glacier risk assessment and water forecasting (preparatory assessments and supporting actions in collaboration with ADB's Glaciers-to-Farms program)	Tajikistan, Turkmenistan	2025
Irrigation, reservoir restoration, and river basin monitoring in Central Asia (coordination with CAREC Water Pillar)	Uzbekistan, Kazakhstan, Turkmenistan	2025
Climate adaptation in mountainous areas (in collaboration with ADB's Glaciers-to-Farms project)	Kyrgyz Republic, Pakistan	2026

ADB = Asian Development Bank, CAREC = Central Asia Regional Economic Cooperation.

[a] Responsible for coordinating the implementation of the regional initiatives and projects, and reporting on progress during working group meetings to the CAREC Climate Change Steering Committee, once established.

Source: ADB.

The glaciers in Central Asia are a critical component of the region's hydrology, serving as the primary source of freshwater for major river systems such as the Amu Darya and Syr Darya, which are essential for the region's agricultural production, hydropower generation, and urban water supply. However, glaciers in Central Asia are under increasing threat from climate change, which has implications for the region's long-term availability and management of water resources.[39] Similar concerns for water security surround the Kura Aras River basin in the South Caucasus, originating in Türkiye and flowing through Georgia and Azerbaijan.

---

[39] E. Zholdosheva et al. 2017. *Mountain Adaptation Outlook Series: Outlook on Climate Change Adaptation in the Central Asian Mountains*. GRID-Arendal.

Glacier retreats and unseasonal glacier melting contribute to unpredictability of water flows and water availability that are set to be further exacerbated by climate change. Transboundary water management and predictable river flows are critical for hydropower generation and the further integration of renewable energy in the regional grid.

Ensuring food security requires addressing the interdependencies between water, energy, and agricultural production. Glacial melt will lead to progressively reduced river flow and increased drought risk. This can have far-reaching consequences for the region's economic, social, and environmental prosperity, particularly for mountain communities and rural women tasked with food production and preparation and thus more vulnerable to climate change.[40]

Regional projects and initiatives under this action plan area will be implemented through and/or in close coordination with ADB's upcoming Glaciers-to-Farms (G2F) regional program (Box 1) and the CAREC Water Pillar Working Group. Energy initiatives focused on mitigation activities for energy efficiency, renewable energy, and grid integration are included under Action Plan Area 3.

BOX 1

Glaciers-to-Farms Regional Program

The Asian Development Bank's Glaciers-to-Farms (G2F) regional program aims to promote climate resilience and low-carbon development focused on climate adaptation and sustainable development in Central and West Asia. G2F envisages mobilizing up to $3.5 billion from the Asian Development Bank and other cofinancing partners (e.g., Green Climate Fund) to invest in projects from 2025 to 2031. It will address the region's vulnerabilities to climate change effects, such as glacier melt, water scarcity, and declining agricultural productivity, which threaten socioeconomic stability.

The proposed G2F program has four components: (i) climate resilience through integrated planning and policy reforms for financial acceleration; (ii) glacier investment and technical facility with pillars on climate-resilient landscapes for water resource management, sustainable agriculture value chains and food security, and enhancing adaptation for climate-vulnerable poor populations in mountainous and rural areas; (iii) establishment of green business financing lines for small and medium-sized agriculture enterprises; and (iv) a G2F regional knowledge-sharing and coordination hub.

The G2F program promotes regional transboundary cooperation and supports the operationalization of the *Climate Action Road Map for Central and West Asia, 2025–2030* and accompanying country climate action plans. Further, it will support countries' implementation of nationally determined contributions and national adaptation plans.

Source: Asian Development Bank.

---

40 United Nations Regional Centre for Preventive Diplomacy for Central Asia. 2014. *Glaciers Melting in Central Asia: Time for Action*. Seminar report. Dushanbe, Tajikistan. 11–12 November.

Regional Glacier Risk Assessment and Water Forecasting

This initiative focuses on conducting a regional glacier risk assessment for improved monitoring and forecasting of regional water resources. The main objective is to assess the impacts of glacier melt and water scarcity on the WEF security nexus and to identify the vulnerability of critical economic sectors, infrastructure, and communities to the effects of glacier melt and water scarcity.

The initiative will support countries' forecasting capabilities for glacier dynamics and water availability based on regional baselines and climate projections. Regional glacier and water resources monitoring could be integrated into the CAREC MHEWS (under Action Plan Area 1) by expanding existing on-site and remote sensing monitoring stations to track glacier mass balance, melt rates and associated water flows, glacial lake outburst floods (GLOFs), and other related hazards. It could also include the creation of a regional atlas of glacier-related natural hazards (GLOFs, mudflows, landslides, avalanches, etc.) on a geographic information system platform integrated into a CAREC climate platform for MHEWS.

Lastly, support will be provided to build countries' technical and institutional capacities for climate-resilient planning focused on the WEF security nexus and enhanced regional cooperation and data-sharing mechanisms for effectively managing shared water resources (linkages with Action Plan Areas 1 and 4).

The regional glacier risk assessment and related initiatives proposed by CAREC members are identified under the G2F program currently being developed by ADB (Box 1). To maximize synergies with CAREC development partners' activities, the CCAP will coordinate with G2F to identify complementarity and reduce potential overlap. For instance, out of the initiatives outlined in these sections, the CCAP could contribute to preparatory assessments and required participatory consultations in collaboration with the G2F design team. The CAREC WGCC and the CAREC climate platform could also provide the venue for coordination on regional elements and knowledge exchange among CAREC countries, while not directly implementing activities that will be included under the G2F program.

Climate Adaptation in Mountainous Areas

The CCAP focuses on responding to the adaptation needs of mountain communities in the CAREC region resulting from changing precipitation patterns, glacier retreats, GLOFs, growing water scarcity, soil erosion, and loss of biodiversity. The needs of these mountainous areas will be addressed in the context of other action plan activities and, where appropriate, with specially targeted actions.

Proposed interventions could include (i) regional initiatives addressing shared challenges of mountain communities and smallholder farmers (especially women-led *dekhan* farms)[41] across CAREC countries, including access to clean and green energy, improving food security and better access to markets, and promoting adoption of climate-smart agriculture practices and crop diversification; (ii) pursuing nature-based solutions (NBS) such as reforestation and restoration of degraded land to reduce natural hazard risk (floods and landslides) while, at the same time, sequestering carbon; and (iii) facilitating the exchange of knowledge and best practices on climate adaptation among mountain communities, local authorities, and relevant regional and national institutions through the CAREC climate platform (linked to Action Plan Area 4).

CAREC will work closely with national and international partners in pursuing these interventions, including the Regional Environmental Centre for Central Asia (CAREC Eco), the Mountain Partnership Secretariat at the Food and Agriculture Organization of the United Nations, and other development partners actively supporting CAREC's mountain areas. CAREC will also assist its member countries in making the case for enhanced attention to the needs of mountain areas in international forums, such as at Conference of the Parties (COP) 29 and future COPs and the International Conference on Glaciers' Preservation in Tajikistan in 2025.

Irrigation, Reservoir Restoration, and River Basin Monitoring in Central Asia

Under the WEF security nexus, the CCAP will seek collaboration with the CAREC Water Pillar to help address the needs and priorities identified by CAREC countries, including upstream climate risk assessments on water and NBS for climate change adaptation to enhance the efficient use of water. While not directly implementing these activities, the WGCC will ensure coordination with the CAREC Water Pillar Working Group, including providing support for knowledge exchange and facilitating regional coordination with other relevant sectors (e.g., energy).

---

[41]   A *dekhan* farm is an individual or family farm in Central Asia.

Panj River near Ishkashim, Gorno-Badakhshan Autonomous Region (GBAO), Tajikistan. The Panj River is fed by numerous glaciers in the high Pamir region and is the headwaters of the Amu Darya; millions of people in Central Asia depend upon it (photo by Matteo Modè).

Collaboration with the CAREC Water Pillar will focus on transboundary projects and investments to increase the resilience of Central Asia's water and energy nexus. It includes a set of proposals under consideration by the CAREC Water Pillar Working Group that demonstrates regional benefits for other countries, while also integrating innovative solutions for climate resilience.[42]

By 2025, three regional projects are expected to be selected by the CAREC Water Pillar Working Group for further development and implementation. Potential projects include the modernization of irrigation systems in the Kyrgyz Republic and Kazakhstan; effective water management of reservoirs in Tajikistan, with downstream benefits for Uzbekistan and Kazakhstan; upgrading of the Kosonsoy water reservoir in the Kyrgyz Republic; and automated monitoring systems and digital transformation for Syr Darya and Amu Darya River basins. The WGCC will work and coordinate closely with the CAREC Water Pillar Working Group to advance the selected regional initiatives.

---

[42]   ADB. Regional: Developing the Central Asia Regional Economic Cooperation Water Pillar. The work of the CAREC Water Pillar started in 2017, with a focus on Central Asia, building on the Almaty Agreement (in the Field of Joint Management of the Use and Conservation of Water Resources of Interstate Courses) and the work of the International Fund for Saving the Aral Sea.

Amu Darya near Termez, Uzbekistan. Regional cooperation is essential for transboundary river management, and monitoring and forecasting of water resources (photo by Matteo Mode).

In addition to water management, the role of agrifood systems, particularly climate-smart agricultural practices such as carbon sequestration through forest, wetland, and soil management, can be integrated to enhance climate resilience further. Agrifood systems, beyond traditional irrigation, play a critical role in climate solutions by improving food security, increasing soil health, and contributing to carbon capture through NBS (link to Action Plan Area 3).

In addition, recognizing the importance of rainfed agriculture in Central Asia's food systems, future regional initiatives could also address the unique needs of lowland areas, where most rainfed crops are grown. Projects that promote sustainable land management practices in these areas—such as soil restoration, carbon sequestration through sustainable agriculture, and ecosystem-based adaptation—would contribute to flood mitigation, reduce dust storms, and enhance biodiversity.

Linkages and Synergies with Existing Initiatives

There are relevant past and ongoing global and regional initiatives from development partners and existing CAREC institutional structures that the proposed interventions under Action Plan Area 2 can build upon and establish linkages with, as indicated in Table 4.

Table 4: Linkages and Synergies for the Water–Energy–Food Security Nexus

Regional Initiative	Project Particulars
Regional Glaciers' Risk Assessment and Water Forecasting	■ The International Conference on Glaciers' Preservation[a] will be hosted in Dushanbe in 2025 to mark the International Year of Glaciers' Preservation as part of the United Nations General Assembly's Glacier resolution that was put forward by Tajikistan.[b] ■ Glaciers-to-Farms regional program ■ Central Asia Regional Economic Cooperation (CAREC) Water Pillar[c]
Climate Adaptation in Mountainous Areas	■ The Kyrgyz Republic development of a road map was supported by the United Nations Development Programme under the 2023–2027 United Nations (UN) General Assembly's Declaration for Five Years of Action for the Development of Mountain Regions.[d] The initiative was a key outcome of the International Year of Sustainable Mountain Development 2022. ■ The Kyrgyz Republic prepared a shared pavilion for mountainous countries at the 29th Conference of the Parties of the UN Framework Convention on Climate Change (UNFCCC), or COP29, dedicated to sustainable mountain development, with support from the Regional Environmental Centre for Central Asia (CAREC Eco), building upon the experience of the COP28's Central Asia pavilion focusing on mountains, climate change, and adaptation.[e] The pavilion will be open and shared by all mountainous countries. ■ Glaciers-to-Farms regional program ■ Deutsche Gesellschaft für Internationale Zusammenarbeit (GIZ) is preparing a regional project, Climate Resilient Landscapes in Central Asia, to strengthen the climate resilience of the most vulnerable populations and the transboundary ecosystems on which they depend and transforming agriculture, forestry, and other land use systems in Central Asia toward low-emission, climate-resilient development. Expected outcomes include increased resilience of land users and rural populations to climate change (through pasture and forest restoration, climate-smart agricultural systems and protected areas, and income diversification opportunities); national climate-resilient planning processes, enabling environment and incentive schemes in place to enable land users to benefit from adaptation measures; and strengthened regional cooperation on climate resilience.[f] ■ The European Bank for Reconstruction and Development is developing a Regional Climate-Smart Agriculture and Resilience Platform for Expedited Transition. This will support access to adaptation finance, knowledge, and technologies for vulnerable farmers, agribusinesses, and local communities through financial institutions (platform expected to launch after 2025).
Irrigation, Reservoir Restoration, and River Basin Monitoring in Central Asia	■ Coordinate with the CAREC Water Pillar Working Group to further develop and implement the selected projects. ■ Food and Agriculture Organization's Central Asia Water and Land Nexus[g]

[a] Centre for Research of Glaciers of the National Academy of Sciences of Tajikistan. *2025—International Year of Glacier Preservation*.
[b] United Nations. 2022. *International Year of Glaciers' Preservation, 2025: Revised Draft Resolution*.
[c] CAREC. *Developing the CAREC Water Pillar*.
[d] For a timeline of the road map, see Zoï Environment Network. *Five Years of Action for the Development of Mountain Regions, 2023–2027*.
[e] CAREC Eco. 2023. *Mountains and Climate Change: Highlighting the Need for Negotiation Group*. Program notes. 6 December; *Central Asia Climate Portal*. 2023. *UNFCCC COP28: Kyrgyzstan Initiative to Establish a Negotiating Group on Mountain Partnership Supported*. 6 December; and UNFCCC. 2023. *Understanding and Closing Adaptation Knowledge Gaps in Mountains and High-Latitude Areas*. *UN Climate Change News*. 11 December.
[f] *Government of Uzbekistan, Ministry of Economy and Finance*. 2024. *A Meeting on the "Climate-Resilient Landscapes in Central Asia" Project Was Held with Representatives of the German Development Agency (GIZ)*. 1 November.
[g] Food and Agriculture Organization of the United Nations. 2024. *Central Asia Water and Land Nexus (CAWLN)*.

Source: Asian Development Bank.

Low-Carbon Growth

Overview

Table 5 summarizes the activities for the low-carbon growth action plan area.

Table 5: Action Plan Area 3—Low-Carbon Growth

OBJECTIVE		
To create the enabling conditions to shift to a low-carbon economy, focused on the energy and transport sectors		
Regional Initiative	**Coordinating Country[a]**	**Indicative Start Date**
Decarbonization of CAREC transport and trade corridors	People's Republic of China, Georgia	2025
Carbon markets development	Kazakhstan, Mongolia	2025
Low-emission zones	Azerbaijan, People's Republic of China, Kazakhstan	2025
Grid readiness and renewable energy integration	Azerbaijan, Georgia, Uzbekistan	2026

CAREC = Central Asia Regional Economic Cooperation.

[a] Responsible for coordinating the implementation of the regional initiatives and projects, and reporting on progress during working group meetings to the CAREC Climate Change Steering Committee, once established.

Source: Asian Development Bank.

The transport sector is a major source of GHG emissions, accounting for 23% of global carbon emissions.[43] Three-quarters of transport emissions are from road transport[44] and the burning of fossil fuels in trucks, car, buses, and trains. In addition, data from the CAREC region shows a steady increase in the yearly growth of transport fleets between 2010 and 2019.[45] While there has been progress in developing electric vehicles, biofuels, and other low-emission technologies, adopting these alternatives has been slow because of the high costs and structural challenges associated with some technologies.[46]

The adoption of renewable energy sources, such as solar, wind, and hydropower, has also been relatively slow in the CAREC region compared to global trends. The CAREC region is heavily reliant on fossil fuels. Some CAREC countries are ranked among the highest per-capita emitters,

---

[43]   P. Jaramillo et al. 2022. Transport. In Intergovernmental Panel on Climate Change. *Climate Change 2022: Mitigation of Climate Change. Contribution of Working Group III to the Sixth Assessment Report of the Intergovernmental Panel on Climate Change.* Cambridge University Press.

[44]   H. Ritchie. 2020. Cars, Planes, Trains: Where Do CO_2 Emissions from Transport Come From? *Our World in Data.* 13 October.

[45]   C. M. Melhuish. 2023. *Decarbonization of the Transport Sector in CAREC Countries: Assessment of Policy Options.* A PowerPoint presentation.

[46]   International Energy Agency. 2023. *Tracking Clean Energy Progress 2023.*

contributing significantly to global GHGs. [47] While all CAREC countries are signatories to the Paris Agreement and have decarbonization goals and targets in their respective NDCs, the share of projects integrating renewable energy in the transport sector is small.

Renewable energy integration requires addressing a range of challenges.[48] Existing electricity grids in the Central Asia region are often outdated and lack the necessary infrastructure to allow for large-scale integration of variable renewable energy sources such as solar. In addition, the grid may not have the required flexibility, transmission capacity, and system control capabilities to manage the fluctuations in renewable energy generation, which can pose risks to grid stability and reliability. Renewable energy projects have upfront capital costs that can be high compared to fossil fuel-based power generation, especially in Central Asia, where there is limited access to affordable financing, particularly for small-scale renewable energy projects, hindering their deployment. Moreover, Central Asia's energy tariffs and subsidies are not yet fully aligned to support the economic viability of renewable energy investments.

Several CAREC countries have extended heating seasons, with a substantial share of energy usage and carbon dioxide emissions attributable to heating. Replacing fossil fuel-based heating systems with renewable alternatives might not always be feasible because of the required investments. On the other hand, energy efficiency represents a largely untapped climate mitigation potential in the CAREC region that could lead to GHG emissions savings and contribute to NDCs targets. The CAREC's Almaty–Bishkek Economic Corridor technical assistance includes support for energy-efficient solutions by switching to efficient heat pumps in Bishkek and developing low-emission zones in Almaty.

Decarbonization of CAREC Transport and Trade Corridors

This initiative will analyze CAREC transport and trade corridors' emissions to help identify required solutions and policy measures to overcome current inefficiencies. CAREC uses the Corridor Performance Monitoring and Measurement tool to assess and track the time and cost of moving goods across borders and along the transport corridors.[49] The tool indicates that border crossing times increased markedly from 12.2 hours in 2019 to 15.1 hours in 2020, primarily because of the lack of integration of national customs systems and inefficient and lengthy border procedures, with implications for travel costs and associated emissions.[50]

The proposed initiative will also identify climate mitigation measures to help cut cross-border times and increase transport efficiency. Potential options include digitalizing trade processes (e.g., integrated paperless custom procedures based on common standardized formats), upgrading cross-border infrastructure to improve border efficiency and reduce transit times, and integrating smart mobility and mass transport into existing transport modalities.

---

[47] M. W. Jones et al. 2024. *National Contributions to Climate Change Due to Historical Emissions of Carbon Dioxide, Methane and Nitrous Oxide.* EU Open Research Repository (accessed in July 2024).

[48] CAREC Secretariat and ADB. 2019. *CAREC Energy Strategy 2030: Common Borders. Common Solutions. Common Energy Future.*

[49] CAREC Secretariat. 2023. *CAREC Corridor Performance Measurement and Monitoring Annual Report 2022.*

[50] CAREC Secretariat and ADB. 2021. *CAREC Corridor Performance Measurement and Monitoring Annual Report 2020: The Coronavirus Disease and Its Impact.*

Windmills along the Almaty–Bishkek road. Renewable energy in CAREC countries has grown significantly, requiring regional cooperation for effective integration in the power grid (photo by ADB).

Grid Readiness and Renewable Energy Integration

This initiative will focus on assessing readiness of the regional grid to increase integration of renewable energy, focusing on both hard and soft infrastructure. The objective is to identify grid instability factors, grid overload, bottlenecks in regional connectivity, and gaps in existing legislation and tariff systems. The assessment will support the identification of required investments for grid modernization and the strengthening of the transmission and distribution infrastructure to accommodate higher levels of renewable energy and achieve enhanced interconnectivity.

ADB's Sermsang Khushig Khundii Solar Project in Mongolia, 2019. Different renewable energy sources need to be integrated effectively to ensure consistent power grid output and stability (photo by ADB).

Other options could include developing and improving legislation to connect CAREC countries better and allow for more efficient regional power trade, such as dynamic tariff systems by implementing flexible electricity pricing to match supply with domestic and regional demand better, ensuring more efficient use of resources.

Lastly, regional cooperation and knowledge sharing will facilitate the exchange of best practices and the development of harmonized regulatory frameworks, and build local technical and institutional capacity through training, capacity-building programs, and the involvement of international experts and organizations. Linkages can be established with Action Plan Area 1 by integrating climate risk assessment into the design of renewable energy projects. For instance, solar power installations in Central Asia are increasingly affected by climate change, such as through extreme heat and the rising incidence of dust storms. In addition, heat waves strain the grid as more and more people resort to air conditioners and refrigerators to cool down. At the same time, dust and sand can significantly reduce the output of solar facilities as service costs are ramped up (e.g., Uzbekistan).[51]

---

[51]   United Nations Convention to Combat Desertification and Food and Agriculture Organization of the United Nations. 2024. *Guideline on the Integration of Sand and Dust Storm Management into Key Policy Areas*.

Low-Emission Zones

People in the CAREC region are exposed to consistently elevated levels of air pollution that are dangerous to human health. Existing data show that average annual particulate matter (PM) 2.5 levels of most capital cities in the region exceed the World Health Organization air quality guidelines.[52] Evidence shows that air pollution is among the most severe environmental risks in the region.[53] Air pollution can lead to several health consequences, such as cardiovascular and respiratory diseases, and adverse effects on the central nervous system, particularly among the most vulnerable.

To reduce emissions and improve urban air quality, this initiative will focus on assessing the impacts of low-emission zones on air quality. This is piloted under the CAREC's Almaty–Bishkek Economic Corridor framework. This initiative will test the establishment of low-emission zones, including alternative transport and heating options to reduce harmful emissions in urban areas.[54] These initiatives can be shared through regional knowledge exchange and replicated in other CAREC countries.

Emissions reduction and energy efficiency can also be achieved via low-carbon infrastructure and the implementation of green building standards for operations in the trade, social, health, and education sectors. The CAREC climate platform can provide the required venue for collaboration and information exchange on green building standards, with policy and implementation varying across the CAREC region.

Carbon Markets Development

Carbon pricing, as part of a broader climate policy architecture, can play an essential role in reducing GHG emissions cost effectively and achieving wider energy system and decarbonization objectives. There is a vast landscape of carbon pricing instruments, including carbon taxes, emissions trading systems, domestic carbon crediting mechanisms, and international carbon markets. Green taxes and reform to fossil fuel subsidies can also serve as the fiscal equivalent of a carbon tax. Carbon pricing can create incentives for energy efficiency and renewable energy investment, which help improve energy security and affordability.[55]

The CCAP will support CAREC countries in the adoption and strengthening of both direct and indirect carbon pricing policies, including carbon taxes, emissions trading systems, international carbon markets (both under Article 6 of the Paris Agreement and the voluntary carbon market), reform to fossil fuel subsidies, and tradable certificates, as aligned with countries' national

---

[52] IQAir. 2023. *World Air Quality Report*. The World Health Organization air quality guideline is 10 micrograms per cubic meter ($\mu g/m^3$), while, in some cases, pollution levels exceed even the more conservative European Union threshold of 25 $\mu g/m^3$.

[53] United Nations Development Programme. 2021. *Tackling Air Pollution in Europe and Central Asia for Improved Health and a Greener Future*.

[54] Other initiatives by the Almaty–Bishkek Economic Corridor (ABEC) include the installation of air quality monitoring stations in the cities of Almaty and Bishkek, with the aim to fill existing knowledge and awareness gaps in air pollution levels of urban areas. See ABEC. *Air Quality*.

[55] ADB. 2022. *Carbon Pricing for Energy Transition and Decarbonization*.

Thick blanket of smog covering Ulaanbaatar. People in the CAREC region are exposed to consistently elevated levels of air pollution that are dangerous to human health (photo by ADB).

circumstances and priorities.[56] CAREC countries have highlighted the lack of technical capacity and incentives to cooperate as the main barriers inhibiting regional cooperation on carbon markets. At the same time, CAREC countries have strong expectations for participation in international carbon markets as net sellers of carbon credits in the global compliance and voluntary carbon markets. However, the level of readiness to engage in international carbon markets varies among countries.

Based on the needs identified by CAREC countries, support under this action plan area could include (i) providing technical and capacity-building support; (ii) promotion of regional knowledge exchange; (iii) integration of carbon markets as part of the broader climate policy architecture; (iv) Article 6 national strategy development; (v) development of carbon baselines to ensure accuracy and consistency of data, particularly in sectors like energy, agriculture, and land use, and robust monitoring, reporting, and verification mechanisms; (vi) carbon asset preparation and monetization support such as for the development of natural climate solutions and NBS projects via bringing in private sector capital; and (vii) exploring approaches that would unlock incentives for cross-border cooperation on climate mitigation.[57]

Training for carbon baseline development could include using and applying carbon capture and storage in specific sector contexts, carbon accounting, and carbon sequestration opportunities within the agriculture and land-use sectors, including knowledge sharing via the CAREC climate platform. This could include establishing standardized carbon quantification methodologies and integrating industrial and NBS such as soil carbon sequestration and reforestation. In this

---

56   On 2–3 September 2024, in Baku, Azerbaijan, CAREC provided an executive training on Article 6 as part of the CCAP quick actions under this Action Plan Area. During the training session, the CAREC WGCC identified carbon markets and pricing as the most suitable types of instruments to support regional cooperation on climate mitigation.
57   CAREC WGCC training on carbon markets. 2–3 September 2024. Baku, Azerbaijan.

context, regional cooperation on knowledge transfer and best practices related to sustainable land management, agroforestry, and climate-smart agriculture would enhance the region's ability to contribute to global carbon capture efforts.

Pilot initiatives focusing on the co-benefits of integrating carbon capture and storage with sustainable land-use practices could demonstrate scalable models for carbon sequestration, aligning with broader goals of climate resilience and rural development (linked to Action Plan Area 2). Additionally, strengthening partnerships and collaborative networks would help mobilize technical expertise and resources to fill the gaps in regional and national carbon baselines, enabling the effective monitoring and verification of carbon capture efforts.

Linkages and Synergies with Existing Initiatives

The proposed interventions under Action Plan Area 3 can build upon and establish linkages with relevant past and ongoing global and regional initiatives from development partners, as well as existing CAREC institutional structures (Table 6).

Table 6: Linkages and Synergies for Low-Carbon Growth

Regional Initiative	Project Particulars
Decarbonization of CAREC Transport and Trade Corridors	- Coordinate with the CAREC Regional Trade Group to develop and implement initiatives to decongest, diversify, and digitalize trade that could contribute to greenhouse gas emissions reduction.[a] Other areas of the group's work that could be linked to the CCAP may include carbon policy instruments as deemed relevant by parties, exploring opportunities in critical minerals,[b] and rapid assessment of the impact of the recent floods on Kazakhstan's trade infrastructure and identification of potential resilience measures.
Grid Readiness and Renewable Energy Integration	- Coordinate with the CAREC Energy Sector Coordinating Committee and its work program and activities such as study tour to Norway and Denmark in June 2024, where energy sector representatives from CAREC countries learned about best practices on grid integration in Europe's Nordic region. - ADB is working with regional and international partners for a scalable Energy Transition Mechanism,[c] initially proposed under the World Economic Forum umbrella in 2018. This is a replicable and scalable market-based model to help accelerate the transition from coal to clean power. - The ADB program in Georgia, Energy Storage and Green Hydrogen Sector Development, includes a project to develop battery storage for further integration of renewable energy into the national grid.[d]
Low-Emission Zones	- Coordinate with the ABEC subcommittee on air quality measurement and establishment of low-emission zones. - The World Bank's air quality work aims to increase regional cooperation and transboundary efforts involving Central Asian countries.[e] - The Kyrgyz Republic has worked with the United Nations Development Programme and the United Nations Environment Programme to improve its air quality monitoring capabilities and data-sharing practices.[f] - Tajikistan has been collaborating with the World Bank and other international organizations to enhance its air quality monitoring and management efforts, including establishing a national air quality monitoring network.[g] - Georgia's experience implementing its road map for air quality monitoring network development is another example of knowledge that could be shared via a CAREC regional platform.[h]

continued on next page

Table 6 *continued*

Carbon Markets Development	■ Various ADB publications show that carbon markets in Asia and the Pacific can help economies reach emissions reduction goals and generate the financing needed for climate action.[i]
	■ UNECE produced a technology brief focusing on CCS and required actions that identified sharing good practices, working via a transboundary approach, and the crucial role of the private sector as the main guiding principles.[j]
	■ ADB supports Pakistan with a technical assistance focused on determining the potential of CCS, looking at energy, agriculture, transport, and industrial processes, with the aim of developing road maps that also include the private sector.[k]

ABEC = Almaty–Bishkek Economic Corridor, ADB = Asian Development Bank, CAREC = Central Asia Regional Economic Cooperation, CCAP = Climate Change Action Plan, CCS = carbon capture and storage, UNECE = United Nations Economic Commission for Europe.

[a] ADB. 2019. *CAREC Integrated Trade Agenda 2030 and Rolling Strategic Action Plan 2018–2020*. This document outlines how CAREC can enhance trade by addressing key challenges, including poor market access, limited economic diversification, and weak institutions.
[b] Critical minerals such as copper, lithium, nickel, and cobalt, and rare earth elements are essential for rapidly growing clean energy technologies, such as wind turbines, electricity grid, solar panels, and electric vehicles.
[c] ADB. Energy Transition Mechanism.
[d] ADB. Georgia: Energy Storage and Green Hydrogen Sector Development Project.
[e] World Bank. 2024. Building a Clean Air Future in Central Asia: The First High-Level Regional Policy Dialogue. Tashkent. 19 June. This event was attended by policymakers and stakeholders from Kazakhstan, the Kyrgyz Republic, Tajikistan, Turkmenistan, and Uzbekistan, together with leading air quality experts from Central Asia and around the world.
[f] United Nations Development Programme and United Nations Environment Programme. 2022. *Air Quality in Bishkek: Assessment of Emission Sources and Road Map for Supporting Air Quality Management*.
[g] World Bank. 2023. *Air Quality Management in Tajikistan*.
[h] N. Megrelishvili. 2021. *Developments of Georgia in Ambient Air Quality Management*. Presentation at UNECE Convention on Long-Range Transboundary Air Pollution. 26 April.
[i] ADB. 2023. *National Strategies for Carbon Markets under the Paris Agreement: Making Informed Policy Choices*.
[j] UNECE. 2022. *Technology Brief: Carbon Capture, Use and Storage*.
[k] ADB. Pakistan: Determining the Potential for Carbon Capture and Storage.

Source: ADB.

CAREC Climate Platform

Overview

Table 7 summarizes the activities for the CAREC climate platform action plan area.

Table 7: Action Plan Area 4—CAREC Climate Platform

OBJECTIVE
To strengthen regional climate cooperation and coordination to address transboundary climate adaptation and mitigation needs and priorities

Regional Initiative	Indicative Start Date[a]
Capacity development on regional climate action	2024
Regional platform for knowledge generation and dissemination	2025
Climate finance and development of bankable regional climate projects	2025

CAREC = Central Asia Regional Economic Cooperation.

[a] The year 2024 is the preparatory phase of the Climate Change Action Plan, including the initiation of some pilot actions in various action plan areas.

Source: Asian Development Bank.

Regional Platform for Knowledge Generation and Dissemination

This initiative aims to establish a CAREC climate platform to strengthen coordination, information, and knowledge sharing on climate action and enhance the capacity of CAREC countries to address climate-related challenges. This is key for implementing the CCAP activities and ensures a harmonized and effective transboundary approach toward climate adaptation, resilience, disaster preparedness, and mitigation activities. The proposed initiative is guided by the CAREC Climate Change Vision (footnote 16), which recommends building an open and inclusive regional institutional platform on climate change.

The CAREC climate platform will facilitate the development and implementation of joint regional climate action initiatives that will require close coordination and cooperation with development partners and CAREC countries to mobilize financial resources and technical expertise for climate-resilient development in the region. The platform will support the collection and dissemination of knowledge and best practices on climate change adaptation and mitigation and facilitate the exchange of experiences and lessons learned among CAREC countries, including knowledge and practices incorporating gender equality perspectives.

The CAREC platform will work in synergy with CAREC clusters to implement the CCAP activities and integrate climate change into the respective cluster strategies and work plans. The CAREC climate platform could also work in partnership with the CAREC Institute, the ADB Institute, the Regional Environmental Centre for Central Asia (CAREC Eco), and other regional institutions to promote the development of a range of knowledge products. Potential activities could include knowledge-sharing events, such as workshops, webinars, and regional forums to facilitate the exchange of experiences and best practices on climate change, collaborative research on climate topics at different sector and thematic levels, and peer-to-peer exchanges among CAREC countries.

CAREC Working Group on Climate Change meeting, Astana, 29–30 May 2024. The working group contributed to the development of the Climate Change Action Plan via a bottom-up participatory approach (photo by ADB).

The CCAP knowledge resources will be disseminated via diverse communication channels and platforms (e.g., CAREC website, social media, newsletters) and translated into CAREC languages to improve accessibility and relevance for national and subnational stakeholders.

Capacity Development on Regional Climate Action

This initiative focuses on providing training support to CAREC countries on climate-related issues across CCAP action areas and key topics to support a more informed and constructive dialogue during United Nations climate change conferences (COP). This includes, among others, capacity building for government officials and policymakers on Article 6 of the Paris Agreement regarding carbon market cooperative mechanisms, to be provided in September and October 2024 by ADB and the United Nations Development Programme in preparation for COP29.

Additional training on Article 6 could focus on regulatory instruments, carbon pricing (via examples and practical applications), and the relevance of baseline development (and distinctions from monitoring, reporting, valuation, and GHG inventory), which could facilitate access and cooperation on carbon market opportunities, emissions trading systems, and carbon credits (from economic and monetary perspectives).

In 2024, ADB supported a training on green public financial management aimed at enhancing CAREC countries' understanding of the consequences of climate change on the management and planning of public sector finance and identifying potential entry points to integrate climate into

Development of the Climate Change Action Plan by the CAREC Working Group on Climate Change, Baku, 2–3 September 2024. The working group identified the Climate Change Action Plan's priority areas of interventions via a participatory approach (photo by ADB).

such management. ADB will support financial cooperation across CAREC and the Association of Southeast Asian Nations (ASEAN) regions to mobilize private sector capital for climate action. This initiative will bring together policymakers, regulators, investors, and private sector leaders from CAREC's Capital Market Regulators Forum and the ASEAN Capital Markets Forum to share knowledge and experiences and deepen cross-regional collaboration on sustainable and green finance.

Training topics will be identified and selected based on needs identified by the Working Group on Climate Change (WGCC) and based on CAREC and COP's respective agendas each year. Support for developing joint initiatives and positions for promoting a One CAREC voice in global and regional climate forums, including COP, will also be provided.

Climate Finance and Development of Bankable Regional Climate Projects

As part of regional cooperation on climate, CAREC will provide technical and capacity-building support to prepare bankable regional climate projects at different stages, such as concept, project proposal design, and technical support for accreditation processes.

Considering the scale of required investments, CAREC is setting up a **Climate and Sustainability Project Preparatory Fund** (CSPPF) that is expected to be endorsed at the CAREC Ministerial Conference on 8 November 2024 and become operational thereafter. The CSPPF will contribute to implementing the CCAP by supporting the preparation of bankable regional projects focused on climate and sustainability objectives.

Linkages and Synergies with Existing Initiatives

Proposed interventions under Action Plan Area 4 can build upon and establish linkages with relevant past and ongoing global and regional initiatives from development partners as well as existing CAREC institutional structures (Table 8).

Table 8: Linkages and Synergies for Regional Climate Cooperation

Regional Initiative	Project Particulars
Regional Platform for Knowledge Generation and Dissemination	■ Coordination with the CAREC Institute and ADBI for climate-related research and knowledge products ■ Coordination with CAREC Eco on regional climate research and experiences (e.g., climate risk assessment, early warning systems, resilience and adaptation in mountainous areas) ■ Collaboration with national and regional research institutions, academic institutions, nongovernmental organizations, and civil society organizations to expand the climate knowledge network and explore joint regional capacity building, training, and learning initiatives

ADBI = Asian Development Bank Institute, CAREC = Central Asia Regional Economic Cooperation, CAREC Eco = Regional Environmental Centre for Central Asia.

Source: Asian Development Bank.

Implementation, Financing, and Monitoring

Institutional Arrangements

The following roles and responsibilities support the successful and effective implementation of the CCAP:

Ministerial level. The CAREC Ministerial Conference provides high-level strategic guidance and endorsement for the implementation of the CCAP. The conference serves as a platform to discuss and debate important policy and strategic issues of regional relevance and exercises overall accountability over the results of the CAREC Program. The CAREC Ministerial Conference is held annually and is attended by ministers of CAREC member countries.

Senior officials level. CAREC senior officials provide executive support and strategic guidance for the CCAP implementation and the integration of climate change with CAREC operations and clusters. The Senior Officials' Meeting (SOM) is the operational-level mechanism of CAREC that monitors progress at the cluster and sector levels and has the authority to consider and endorse complex multicountry and multisector projects. The CAREC SOM also serves as a mechanism to ensure the effective implementation of the policy and strategic decisions made by the Ministerial Conference. The CAREC SOM, composed of senior-level officials from the relevant agency in overall planning or finance, is convened annually to prepare for the CAREC Ministerial Conference.

CAREC national focal points. Each country has a senior official appointed as CAREC national focal point (NFP) to ensure effective coordination among all relevant government agencies and development partners at the country level in matters related to regional cooperation. CAREC NFPs will support the WGCC in coordinating with sector agencies to develop and implement CCAP regional initiatives and projects.

Climate Change Steering Committee. A steering committee may be established to oversee and guide CAREC's efforts on climate action and provide strategic guidance to the WGCC on CCAP implementation, including ensuring alignment of sector strategies, action plans, and project portfolios across clusters with the CAREC Climate Change Vision. The steering committee is proposed to be composed of CAREC NFPs and high-level officials from member countries' environmental, ecology, and climate change ministries. The committee will report annually to the CAREC Ministerial Conference on the progress of the CAREC climate agenda.

CAREC Working Group on Climate Change. The CAREC WGCC plays a vital role in the CCAP development and implementation.[58] Its responsibilities include (i) coordinating with and supporting sector committees and working groups for integrating climate change with CAREC 2030's operational areas; (ii) identifying priority sectors and regional initiatives on climate adaptation and mitigation, in alignment with countries' NDCs and climate strategies and plans; (iii) developing the CAREC climate project portfolio and identifying potential financing sources; (iv) building multistakeholder consensus and public–private partnerships, and sharing best practices for regional climate action; and (v) promoting a One CAREC Voice on the climate change agenda at global and regional forums such as the COPs. The WGCC comprises representatives of CAREC member countries, development partners, and the CAREC Secretariat. The WGCC will annually report to the CAREC SOM and the Climate Change Steering Committee on the CCAP implementation. Each year, the CAREC host country (rotating annually) will chair the WGCC meetings, with support from the CAREC Secretariat (the terms of reference for the WGCC are described in Appendix 1).

Coordinating countries. They will update the WGCC on the implementation status of the regional activities they are responsible for coordinating. Coordinating countries will actively establish linkages and synergies with other working groups, such as the water pillar, transport, and energy, by participating in relevant working groups' meetings and reporting back to the WGCC, and vice versa. Coordinating countries will inform the WGCC of the progress of initiatives included in the CCAP but implemented under other working groups.

The CAREC Secretariat. It supports the WGCC in conceptualizing, designing, and implementing the CAREC CCAP and its activities. This includes organizational support for WGCC meetings and coordination and sharing of information with CAREC member countries, development partners, CAREC sector committees and working groups, and other regional and national stakeholders.

CAREC sector committees and working groups. The WGCC will coordinate, link, and establish synergies with committees and working groups across CAREC sectors to advance the implementation of CCAP initiatives. Most of the proposed regional initiatives and projects in the CCAP require sector committees and working groups to be actively involved and/or possibly take the lead regarding implementation. Focus sector group sessions and/or cross-sector meetings of the WGCC with other sector committees and working groups will be organized to raise awareness and advance the implementation of the CCAP initiatives through a coordinated approach.

---

[58] The WGCC is primarily composed of senior- and technical-level government officials from CAREC countries' environmental and climate change ministries and agencies, and other ministries. In 2024, the WGCC met four times (twice virtually and twice in-person in Astana, Kazakhstan and Baku, Azerbaijan).

Development partners. These partners participate at several levels in the overall strategic formulation and implementation of the CCAP and further integration of climate change in the CAREC portfolio, including as active members of the WGCC. To optimize resources, development partners could be assigned to take the lead in implementing various CCAP initiatives and projects based on their core expertise and ongoing initiatives.

CAREC Institute. The institute will play a key role in generating and disseminating climate-related knowledge to support the various regional initiatives and projects of the CCAP under the guidance of and in close coordination with the WGCC and the CAREC Secretariat. The CAREC Institute will also be responsible for establishing collaboration and partnerships with global and regional climate research institutions to ensure access to the latest knowledge and best practices on climate adaptation and mitigation solutions.

Resource Mobilization

The implementation of CCAP initiatives and projects requires considerable incremental financial resources.[59] An analysis of the CAREC project portfolio between December 2016 and December 2023 shows that only 3% were invested in climate adaptation while just 5% were invested in climate mitigation.

To meet the substantial financing requirements and support the CCAP implementation, climate financing for CAREC projects must be scaled up through new financing modalities and solutions, and coordinated from various sources. A wide range of sources will have to be tapped, including domestic resource mobilization, in particular, revenues generated by the effective management of energy subsidies and carbon pricing (multilateral carbon market solutions) that will support the energy transition.[60] Another potential source is private sector finance, which includes green bonds, international reinsurance, and capital markets for regional risk transfer solutions.

However, developing domestic resource bases and private sector financing will take time and will also be challenging. External official funders will, therefore, be a critical source of climate finance for the CAREC region in the foreseeable future. This includes bilateral and multilateral development partners' financing, such as via specific climate funds (e.g., the Green Climate Fund, Global Environment Fund, Adaptation Fund, Systematic Observations Financing Facility, and the Loss and Damage Fund).

The official financing architecture is fragmented and lacks coordination; each funding agency has its own priorities, conditions, and procedures, so official finance can be challenging to access and manage. Moreover, most funders are focusing on national rather than regional climate action. Finally, there is a dearth of bankable public and private investment projects supporting climate action in the CAREC region.

---

[59] The CAREC Climate Change Scoping Study provided a rough estimate of $26 billion in total climate action financing requirements for CAREC (excluding the People's Republic of China), of which $10 billion could be raised domestically and $16 billion internationally.

[60] G. Verdier et al. 2022. Revenue Mobilization for a Resilient and Inclusive Recovery in the Middle East and Central Asia. *IMF Departmental Paper.* No 2022/013. International Monetary Fund.

In this context, CAREC, in cooperation with the CAREC Institute, can play a vital role in supporting domestic resource mobilization and private financing efforts and in helping to facilitate access to official finance, especially for regional projects. This involves supporting (i) the development of national carbon pricing policies and institutions; (ii) capacity building for accessing private sources of climate finance; (iii) information sharing, coordination, and cofinancing among development partners for regional investment projects; (iv) developing the capacity of national authorities to access international climate funds; and (v) strengthening the climate project preparation capacity of CAREC countries.

Multilateral development institutions, such as ADB and the World Bank, could play a lead financing role in this context. ADB has outlined climate change mitigation, adaptation, and environmental sustainability as central pillars in its Strategy 2030, aiming to reach $100 billion in climate finance by 2030. Of this, $34 billion will be earmarked specifically for adaptation and resilience initiatives, presenting opportunities for ADB to fill CCAP financing needs.[61]

The World Bank has committed to align all its financing operations with the goals of the Paris Agreement, while setting targets of 45% of annual financing to fund climate-related projects, with a 50% split between climate mitigation and adaptation.[62] The World Bank commitment overall to mainstreaming climate change into Central Asia is via a regional portfolio of $12 billion.

Similarly, the Asian Infrastructure Investment Bank (AIIB) has ambitious climate financing targets, committing to allocate at least 50% of its annual financing approvals as climate finance by 2025. AIIB's thematic priorities align with the CAREC Climate Change Vision (i.e., green infrastructure, connectivity and regional cooperation, and private capital mobilization). AIIB has climate-related projects and operations in CAREC countries. It is committed to delivering impact, including contributing to the CCAP implementation and supporting needed policy reform.

ADB is also supporting the Kyrgyz Republic by establishing a Climate Finance Center to address climate finance needs.[63] The center aims to act as a one-stop shop at the country level to help coordinate financial assistance available through regional and international climate funds. This example could be replicated in other CAREC countries.

Crucially, to strengthen project preparation capacities of CAREC countries and scale up regional climate finance, CAREC is establishing a dedicated financing mechanism, the Climate and Sustainability Project Preparatory Fund (CSPPF), expected to be launched at the CAREC Ministerial Conference in November 2024. The CSPPF will help address and narrow down CAREC countries' financing gaps in achieving climate commitments and the United Nations Sustainable Development Goals by supporting the preparation of bankable regional projects focused on climate and sustainability objectives consistent with CAREC countries' Paris Agreement and NDCs targets.

---

61   ADB. 2024. ADB Commits Record Climate Finance of Almost $10 Billion in 2023. News release. 31 January.
62   World Bank Group. 2021. *World Bank Group Climate Change Action Plan 2021–2025: Supporting Green, Resilient, and Inclusive Development.*
63   ADB. *Regional: Enabling Green Recovery in Central and West Asia through a Sustainable Financing Program.*

The CSPPF is intended for regional projects (i.e., projects that promote collaboration between two or more CAREC countries and are expected to generate economic benefits for more than one country in the CAREC region). The CSPPF can play a crucial role in advancing the implementation of the CCAP through technical assistance to conduct project preparatory activities, including prefeasibility and feasibility studies and small-sized grant components of regional climate projects that demonstrate and/or pilot innovative features and adaptation and mitigation co-benefits. CAREC could take into consideration ADB's experience in the development of the ASEAN Catalytic Green Finance Facility (Box 2).

BOX 2

Association of Southeast Asian Nations' Catalytic Green Finance Facility

The Central Asia Regional Economic Cooperation (CAREC) Program could take into consideration the Asian Development Bank's experience in the development of the Association of Southeast Asian Nations (ASEAN) Catalytic Green Finance Facility (ACGF), supporting governments in Southeast Asia to prepare and finance infrastructure projects that promote environmental sustainability and contribute to climate change goals. To be financed under the ACGF, projects must be sovereign or sovereign-guaranteed and fulfill eligibility criteria, which include clear environmental goals and targets, a financial sustainability plan, and a road map for attracting private capital investment. In addition to the success in cofinancing and resource mobilization, the ACGF experience can serve as a guiding example for CAREC in terms of establishing knowledge partnerships (as the facility has in-kind knowledge partnerships with Climate Bonds Initiative, Global Green Growth Institute, Infrastructure Asia, and the Organisation for Economic Co-operation and Development).

Source: Asian Development Bank.

The CSPPF will be administered by ADB, with contributions from bilateral and multilateral donors and other sources as financing partners. A steering committee for the CSPPF will be chaired by the director general of ADB's Central and West Asia Department. ADB's Regional Cooperation and Integration Unit will be the fund manager. Financing partners will hold regular consultation meetings to deliberate on CSPPF decisions and strategic directions. The steering committee will consult with the WGCC on CAREC countries' needs and priorities for climate and sustainability projects for funding under the CSPPF.

Monitoring and Evaluation

A set of measurable indicators has been developed to track progress and evaluate the effectiveness of the CCAP's implementation (Appendix 2). During the implementation of the CCAP, indicators may be adjusted and baseline values for each indicator will be developed. Roles, responsibilities, and resource requirements should be clearly established for the effective monitoring and evaluation of the CCAP implementation, including responsible parties and mechanisms for regularly collecting, analyzing, and reporting on the indicator data.

As part of the monitoring and evaluation requirements, the WGCC could provide a venue for regular consultation and feedback on the CCAP implementation performance. This could be done in close coordination with CAREC NFPs, sector committees, working groups, and other relevant stakeholders involved in the monitoring and evaluation process to enhance the relevance and ownership of the CCAP.

Launch of the Climate Change Action Plan at a high-level session on the CAREC Partnership for Climate, Innovation, and Trade at COP29, 14 November 2024, in Baku, Azerbaijan. The launch marked the culmination of CAREC countries' efforts in 2024 for the development of the Climate Change Action Plan (Photo by ADB).

Appendix 1

Terms of Reference of the CAREC Working Group on Climate Change

Background

The Central Asia Regional Economic Cooperation (CAREC) member countries face significant impacts from climate change, which are expected to escalate further, as demonstrated in the Climate Change Scoping Study. Addressing the causes and effects of climate change in the region is of the highest urgency. It will require prompt and effective national and regional responses. CAREC is well placed as a platform to coordinate climate action in the region.

The 22nd CAREC Ministerial Conference in November 2023 adopted the CAREC Climate Change Vision. This vision represents an ambitious agenda designed to ensure adequate regional climate action by CAREC member countries and development partners from 2024 to 2030.

Role and Functions

The role of the CAREC Working Group on Climate Change (WGCC) is to support the efforts on climate action of the CAREC Climate Change Committee (CCSC), with technical advice and substantive inputs in the following areas of the CCSC's responsibility, including decisions regarding the appropriate sequencing of the various activities to be pursued:

- develop the CAREC Climate Change Action Plan (CCAP);
- develop the CAREC climate project portfolio;
- ensure that the CCAP and climate project portfolio are aligned with countries' nationally determined contributions' targets and national climate strategies and plans for adaptation and mitigation of CAREC countries, as applicable;
- identify and develop specific regional initiatives to strengthen the enabling environment for climate action by harmonizing the legal and regulatory environment and by building capacity;
- build and maintain a multistakeholder consensus across the region;
- gather and share best practices for regional climate action;
- develop and launch a strategic communications plan for the CCAP;

- work with development partners to seek and secure project funding for regional climate projects and avoid duplication of efforts;
- build public–private partnerships for regional climate action;
- annually report to the CAREC Ministerial Conference on the implementation progress of the CCAP;
- review and update the CAREC Climate Vision document and the CCAP every 3 years to adapt to changing conditions and needs; and
- convene the WGCC to meet at least twice per year. During 2024, the CAREC Secretariat will chair the WGCC, after which the CAREC host country representative will chair the meetings.

The CAREC Secretariat will provide organizational support for the WGCC and prepare draft documents for its review, input, and, where appropriate, endorsement.

Appendix 2

Climate Change Action Plan Indicators

Action Plan Area	Objective / Outcome	Indicators
Climate risk, preparedness, and health	Resilience of regional infrastructure and communities, and countries' capacities to prepare for and respond to climate and disaster risk increased	■ Number of adaptation projects being conceptualized resulting from regional climate risk assessments ■ Number of Central Asia Regional Economic Cooperation (CAREC) countries exchanging early warning information and data, including for extreme weather events ■ Number of CAREC countries establishing policies and measures for climate change health adaptation and preparedness, including regional actions ■ Number of disaster risk financing instruments piloted in CAREC countries
Water–energy–food security nexus	Countries' resilience to climate shocks across the water, agriculture, and energy sectors improved	■ Number of CAREC countries with improved capacity for forecasting glacier dynamics and water availability ■ Number of enhanced monitoring solutions piloted in transboundary river basins ■ Number of people in mountainous areas, including smallholder farmers, with improved capacity for climate-smart agriculture practices
Low-carbon growth	Enabling conditions to shift to a low-carbon economy promoted	■ Emissions reduction through improved border-crossing procedures and digitalization of trade processes ■ Emissions reduction resulting from regional renewable energy projects ■ Number of low-emission zones piloted in CAREC countries
CAREC climate platform	Regional climate cooperation and coordination to address transboundary climate adaptation and mitigation needs and priorities strengthened	■ Number of public and private stakeholders (sex-disaggregated) trained on key climate-related topics ■ Amount of financing resources mobilized for regional climate projects in CAREC countries

Source: Asian Development Bank.

www.ingramcontent.com/pod-product-compliance
Lightning Source LLC
LaVergne TN
LVHW071455180726
843512LV00018B/1386